Flavored Oils and Vinegars

Flavored Oils and Vinegars

Love Food™ is an imprint of Parragon Books Ltd

Parragon
Queen Street House
4 Queen Street
Bath BA1 1HE, UK

Copyright © Parragon Books Ltd 2007

Love Food™ and the accompanying heart device is a trademark of Parragon Books Ltd.

ISBN 978-1-4054-9570-7

Printed in China

Text by Ann Kleinberg
Produced by the Buenavista Studio s.l.
Photography by Günter Beer
Home Economy by Stevan Paul
Design by Cammaert & Eberhardt

Notes for the reader

This book uses imperial, metric, and U.S. cup measurements. Follow the same units of measurement throughout; do not mix imperial and metric. All spoon measurements are level: teaspoons are assumed to be 5 ml, and tablespoons are assumed to be 15 ml. Unless otherwise stated, milk is assumed to be whole, eggs and vegetables such as potatoes are medium, and pepper is freshly ground black pepper. Recipes using raw or very lightly cooked eggs should not be served to infants, the elderly, pregnant women, convalescents, and anyone suffering from an illness. The times given are approximate only.

Contents

Introduction

Oils and vinegars have appeared on our tables for thousands of years. Evidence has been found in ancient Greek and Egyptian writings, biblical texts, and archaeological finds to suggest that early civilizations appreciated the taste and versatility of these delicious ingredients. How wonderful that something so ancient and venerated is still an integral part of our cuisine today.

In addition to its importance as a cooking medium, oil serves as a preservative (think sun-dried tomatoes, sardines, and olives) and as the basis for most dressings. One might say that oil is the soul of a salad—providing a foundation for all the other ingredients. Is there anything more perfect than a lovely green salad dressed with a fine olive oil and a sprinkle of fresh lemon juice?

Vinegar came to us by accident—and what a happy accident it was! A vat of grape juice that was intended for wine was probably left open or fermented too long, and the result, known in French as "vinaigre" or "sour wine", was the mother of what has now become a vast vinegar industry.

Suddenly the world is taking notice of oils and vinegars—and it is about time! You can enjoy them just as they are, but why not be a bit experimental? Infused with herbs, spices, nuts, or fruit, oils and vinegars take on a whole new life—adding a wonderful flavor dimension. The essence of the flavor soaks into the dish and upgrades it to something a little more special.

The versatility of flavored oils and vinegars never ceases to amaze, and this book will provide the basis for wonderful recipes to inspire and delight. The process is so simple: follow the guidelines, use caution, and have fun. Use the recipes presented here or let your imagination run wild and flavor with your own combinations.

Introduction to Oils

Edible oil has a rich history shared by many cultures over many centuries. It is used as a cooking medium, a preservative, a medicine, and best of all—food. The Chinese and Japanese developed methods to extract oil from soy plants; Mediterranean people used olives; Mexicans and North Americans were fond of peanuts and sunflower seeds; and in Africa, palms and coconuts provided the basis for oil. Other sources of oil are cotton, safflower seeds, watermelon seeds, grape seeds, rapeseed, corn, nuts, avocados, and, of course, animals.

Thanks to the popularity of the Mediterranean diet, olive oil has captured the world's attention with its health-inducing benefits. Other beneficial oils, such as canola, are almost flavorless and present the perfect backdrop for flavorful additions.

There are just a few guidelines to follow before getting started. The most important to remember is this: oil improperly stored can encourage the growth of bacteria. When herbs or vegetables with high water content (like garlic) are mixed with oil and stored in a non-refrigerated place, an oxygen-free environment is created that can lead to botulism. Careful preparation and refrigerated storage will prevent problems.

The following points are important to remember:

1. Invest in some good glass jars and sterilize them (pour boiling water over them and let them stand for 10 minutes). Do not use cheap jars—the metal ring may rust and the rubber seals may disintegrate.
2. Canola oil is a perfect base for most flavorings because it does not have a distinctive flavor. Olive oil is great for the heartier herbs that can handle its strong flavor.
3. Bruising or crumbling herbs before adding them to the oil releases their flavor.
4. Heating the oil will help infuse the flavor of the added ingredients.
5. Make small amounts and always store flavored oils in the refrigerator.
6. Remove from the refrigerator and bring to room temperature before using.

Basil Oil

The concept of flavored oils must have been designed with basil in mind. There are so many recipes for basil and oil—why not combine them and be ahead of the game?

2 cups fresh basil leaves

2 cloves garlic, halved

1 cup olive oil

Wash the basil leaves and dry them well. Prepare a bowl of ice water.

Bring a saucepan of water to a boil, add the basil leaves, and blanch for five seconds. Scoop out the leaves and plunge immediately into the ice water to stop the cooking process. Drain out all the water and squeeze the leaves to get rid of as much of the water as possible. Dry them between layers of paper towels. Chop coarsely and place in a clean jar. Add the garlic.

Gently heat the oil over low heat until warmed and fragrant—about five minutes. Be sure that it does not boil or burn. Remove from the heat and pour the oil into a clean jar over the basil leaves. Let cool, cover, and store in the refrigerator. Strain out the basil within a week.

Note: The olive oil has a very dominant flavor. If you prefer a stronger basil flavor, use canola oil instead.

Mint Oil

You will look for excuses to use this wonderful oil—it is so refreshing and, well, minty!

2 cups mint leaves

1 cup canola oil

Wash and dry the mint leaves. Prepare a bowl of ice water.

Bring a saucepan of water to a boil, add the mint leaves, making sure that they are all submerged, and blanch for five seconds. Scoop out the leaves using a slotted spoon and plunge immediately into the ice water. Drain, squeezing out all the water from the leaves. Dry the leaves between layers of paper towels.

Coarsely chop the leaves and add to a blender or food processor. Pour in some of the oil and process. Add more oil, continuing to process until all the oil is added. Scrape the blender or processor with a rubber spatula, and process again, making sure that all the leaves are finely minced and well combined with the oil.

Strain through a cheesecloth and pour into a clean jar. Refrigerate until ready to use. Bring to room temperature before use.

Tomato, Mozzarella, and Basil Kabobs

12 cherry or plum tomatoes

12 fresh basil leaves, or more

12 baby mozzarella balls

basil oil

freshly ground black pepper

Makes 12 skewers

Making this tasty dish is easy. Just put together the ingredients (and add any of your own), skewer them, and serve to your delighted guests.

Cut the tomatoes in half crosswise. Wash and dry the basil leaves.

Thread each skewer with a tomato half, a basil leaf, and a baby mozzarella ball. Add another tomato half to the end.

Arrange the skewers on a serving platter, drizzle them with the basil oil, sprinkle with a bit of the black pepper, and serve.

Bruschetta with Tomato, Red Onion, and Basil Salsa

1 large baguette or focaccia loaf

2 tbsp **basil oil**

basil salsa

10 plum tomatoes

2 red onions

$^1/_2$ cup basil leaves (green or purple)

juice of 2 lemons

salt and freshly ground black pepper

Serves 6

This is a wonderful idea for a Mediterranean-style lunch or snack. Serve it alongside a simple salad with a chilled bottle of white wine.

Preheat the oven to 450°F/235°C.

Slice open the baguette. Place on a baking sheet, brush with some of the basil oil, place in the oven, and toast until golden.

To make the salsa, blanch the tomatoes in boiling water for about five seconds to allow for the easy removal of the skin. Plunge the tomatoes into some ice water. Peel, slice in half, seed, and dice. Chop the red onions and basil leaves.

Combine the tomatoes, onions, and basil with the lemon juice and the remaining basil oil. Season with salt and pepper to taste. Spoon the salsa over each slice of toasted bread and serve.

Yogurt Soup with Pomegranates and Mint Oil

4 cups plain yogurt

3 whole cucumbers, peeled and diced

4 cloves garlic, minced

3 tbsp fresh dill, finely chopped, plus extra to garnish

3 tbsp fresh mint leaves, finely chopped

juice of $1/2$ lemon

salt

white pepper

$1/8$ cup pomegranate seeds

mint oil

mint oil ice cubes

Serves 4

This soup is so easy to make, so refreshing to drink, and tastes absolutely delicious.

In a large mixing bowl, combine the yogurt, cucumbers, garlic, dill, mint leaves, and lemon juice. Season with salt and pepper to taste. Chill for several hours.

Pour the yogurt soup into soup bowls, garnish with the pomegranate seeds, chopped dill, and a drizzle of the mint oil. If you have them on hand, add a few mint oil ice cubes.

Note: Prepare the mint oil ice cubes in advance by pouring the mint oil into ice cube trays. Freeze and use whenever desired.

Garlic, Chili, and Oregano Oil

Put some of this deliciously spicy oil on a plate, serve with warm, crusty bread, some cheese, and olives, and you will have a Mediterranean feast!

5 cloves garlic

2 tbsp red hot chile, seeded

1 tsp dried oregano

1 cup canola oil

Preheat the oven to 300°F/150°C.

Cut the garlic cloves in half lengthwise. Using gloves, remove the seeds from the chile pepper and chop the pepper into small pieces equaling two tablespoons.

Combine the garlic, peppers, and oregano with the oil in a 2-cup glass measuring cup (make sure it is ovenproof). Place on a glass pie plate in the center of the oven and heat for 1½-2 hours. The temperature of the oil should reach 250°F/120°C if you are using a digital thermometer.

Remove from the oven, let cool, and strain through cheesecloth into a clean jar. Store covered in the refrigerator. You can also leave the garlic and pepper pieces in the oil and strain before using.

Parsley and Cilantro Oil

This oil is so fresh tasting and green! It can be sprinkled on soup or used as a garnish for salads. Feel free to change the ratio of leaves if you prefer one herb over the other.

$^1/_2$ cup fresh parsley leaves

$^1/_2$ cup fresh cilantro leaves

1 cup canola oil

Wash and drain the leaves. Prepare a pot of water, bring to a boil, and submerge the leaves. Blanch for five seconds. Drain the leaves and dry well.

Heat the oil in a saucepan, bring to a near-boil and let simmer for one to two minutes.

Combine the warmed oil and leaves in a blender bowl or food processor. Process until well combined.

Pour through cheesecloth and strain into a clean jar. Cover and refrigerate.

Roasted Antipasto with Garlic, Chili, and Oregano Oil

1 medium sweet potato, peeled and cut into chunks

2 fennel bulbs, quartered

2 leeks, white part only, cut diagonally into two-inch pieces

1 head garlic, cut in half horizontally

10 shallots, peeled

½ cup **garlic, chili, and oregano oil**

2 tbsp coarse salt

Serves 6

These vegetables are only a suggestion. Choose any seasonal vegetables, sprinkle with the wonderfully aromatic garlic, chili, and oregano oil, and roast away!

Preheat the oven to 450°F/235°C.

Arrange all the vegetables on a baking pan or silicone sheet. Brush with the garlic, chili, and oregano oil and sprinkle with salt. Place in the oven and roast for 20 to 25 minutes. Some of the vegetables may soften sooner than others, so keep an eye on them and remove those first.

Remove the tray from the oven, brush the vegetables with a little more of the flavored oil, and serve.

Focaccia with Tomato, Mozzarella, and Red Onion

Focaccia

1 tsp active dry yeast

1 cup warm water

1 tbsp sugar

2$\frac{1}{2}$ cups all-purpose flour, plus extra for dusting

1 tsp salt

4 tbsp olive oil

Topping

Garlic, chili, and oregano oil

4 plum tomatoes, thinly sliced

1 red onion, thinly sliced

$\frac{1}{2}$ tsp coarse (kosher) salt

7 oz/200 g mozzarella, sliced

$\frac{1}{4}$ cup fresh basil leaves, torn

Makes 1 large loaf

Any combination of ingredients can be used for the toppings, but these work beautifully with the delicious garlic, chili, and oregano oil.

In a large mixing bowl, combine the yeast, $\frac{1}{2}$ cup of the warm water, the sugar, and one tablespoon of the flour. Let stand for about 10 minutes until it bubbles and becomes foamy.

Add the remaining water and flour and the salt. Using a dough hook, mix all the ingredients together. Gradually pour in the oil and continue to mix until a ball is formed. Cover the bowl with plastic wrap, set it in a warm place, and let the dough rise for 1 hour. It should double in size.

Preheat the oven to 450-500°F/235-260°C. Grease a baking sheet and sprinkle a little flour on it. Punch down the ball of dough and, using oiled hands, spread it out over the entire baking sheet. Let it rise again for 10 minutes.

Using your fingertips or the back of a spoon, punch dimples into the dough.

Brush some of the garlic, chili, and oregano oil onto the dough, and cover it with slices of tomato and onion. Sprinkle with the coarse salt and pour on a little more of the flavored oil. Bake for 15 minutes.

Remove from the oven, add the slices of mozzarella, and return it to the oven for another 5 minutes or until the cheese melts.

Remove the focaccia from the oven, sprinkle with torn basil leaves, cut into squares, and serve.

Tuna Ceviche with Parsley and Cilantro Oil

7 oz/200 g red tuna fillet
(about one serving)

1 medium red onion, diced

1 red bell pepper, diced

4 tbsp lemon juice

1 lemon, grated zest only

3 tbsp orange juice

$1/3$ cup **parsley and cilantro oil**

$1/4$ tsp chopped ginger

$1/2$ tsp chopped chile

2 tbsp balsamic vinegar

salt and freshly ground black pepper

flat-leaf parsley leaves, to garnish

Serves 2
Ceviche is not difficult to make, as this recipe will demonstrate, and the result will prove that somethings are best when kept simple!

Chop the tuna into very small pieces. Combine with the remaining ingredients, mix well, cover and chill for at least 1 hour in the refrigerator before serving.
Serve with crackers and add a few parsley leaves, to garnish.

Roasted Tomato Oil

This versatile oil tastes divine on almost anything—try it on pizza, pasta, or focaccia for a tasty treat.

4-6 plum tomatoes

1 cup canola oil

Preheat the oven to 400°F/200°C.

Thinly slice the tomatoes and place them on a lightly greased baking sheet. Place in the oven and roast until they start to char. Remove from the oven and let cool.

Heat the oil in a saucepan. Bring to a near-boil and let it simmer for one to two minutes.

Combine the tomatoes with the warmed oil, and process in a blender or food processor. Process until the tomatoes are well incorporated into the oil. Strain through cheesecloth and pour into a clean jar. Refrigerate.

Lemon Pepper Oil

Pasta, chicken, and fish complement the flavors of this oil particularly well, but it is a good staple to keep in the kitchen, as it works well with almost anything.

zest of 1 lemon

1 whole lemon

2 tsp multicolored peppercorns

1 cup olive oil

Prepare a double boiler. Bring the water in the bottom pot to a boil, lower the heat, and simmer.

Cut the lemon zest into thin strips, making sure not to include the white pith. Thinly slice the other lemon. Crush the peppercorns using a mortar and pestle. Put the strips of lemon zest, lemon slices, peppercorns, and olive oil in the top of the double boiler.

Cook over simmering water for 1 hour. The oil should reach a temperature of 250°F/120°C, tested with a digital or candy thermometer. Make sure not to let the oil burn.

Remove from the heat, let cool, and strain through cheesecloth into a clean jar. Cover and store in the refrigerator. You can also leave the lemon strips and pepper in the jar and refrigerate, and then strain before using.

Quinoa Salad with Sun-Dried Tomatoes, Black Olives, and Feta

Salad

1 cup quinoa

2 cups water

10 sun-dried tomatoes (in oil, drained)

$1/2$ cup feta cheese, crumbled

2 scallions, white part, chopped

$1/3$ cup mixed fresh herbs (basil, parsley, cilantro), chopped

$1/4$ cup pitted black olives, chopped

Dressing

$1/3$ cup **roasted tomato oil**

3 tbsp fresh lemon juice

1 clove garlic, crushed

salt and freshly ground black pepper

Serves 4

This salad is perfect as a main course or as a side dish for chicken or fish. It is delicious and colorful and really good for you!

Spread the quinoa on a dish and pick out any pieces of grit. Rinse the grains thoroughly in a fine-mesh sieve and drain.

In a medium saucepan, bring the water to a boil over high heat, stir in the quinoa, and return to a boil. Lower the heat, cover, and simmer for about 15 minutes or until all the liquid has been absorbed. Remove from the heat, fluff up the quinoa with a fork, and transfer to a bowl. Let cool a bit.

Add the remaining ingredients to the bowl and mix with the quinoa.

Whisk the dressing ingredients together, pour over the quinoa, toss, and serve.

Fettuccine with Lemon Pepper Seafood

$^1/_3$ cup **lemon pepper oil**, plus extra for serving

6 cloves garlic, crushed

$1^1/_2$ lbs/675 g mixed seafood (shrimp, calamari, mussels)

dash of vodka

$^1/_2$ cup white wine

1 sprig tarragon, leaves only

dash of salt

1 lb/450 g fettuccine

flat-leaf parsley, chopped, to garnish

Serves 4

This is a great dinner-party dish, as it can be prepared in minutes, but your guests will never know from the delicious taste!

Heat a wok or deep skillet and add the lemon pepper oil. When the oil is hot, add the garlic and seafood. Stir for one minute. Add a dash of vodka, the white wine, tarragon leaves, and salt. Keep stirring until the seafood is cooked through.

　　Prepare the fettuccine according to the directions on the package. When the fettuccine is cooked, drain and add to the seafood mixture. Toss well and serve immediately on warmed plates. Drizzle more lemon pepper oil on top to serve and garnish with the chopped parsley.

Rosemary, Lemon, and Thyme Oil

You may want to make this just for the glorious aroma that permeates the house as the oil is heating.

5 sprigs rosemary
(each about 5 inches long)

10 to 15 sprigs thyme
(each about 5 inches long)

zest of 2 lemons

1 cup canola oil

Preheat the oven to 300°F/150°C.

Remove the leaves from the rosemary and thyme sprigs. Cut the lemon zest into strips.

Pour the oil into a two-cup ovenproof glass measuring cup and add the leaves and lemon zest strips. Place the cup on a pie plate in the center of the oven and heat for one and a half to two hours.

If you have a digital thermometer, test the oil. It should reach a temperature of 250°F/120°C before you remove it from the oven. Let cool for at least 30 minutes.

Store in the refrigerator as is, or strain through cheesecloth and refrigerate.

Lemongrass and Lime Oil

Lemongrass is a treasured ingredient in Southeast Asian cuisine and tastes even better when paired with lime.

10 stalks lemongrass

zest of 1 lime

$1/2$ lime

1 cup canola oil

Preheat the oven to 300°F/150°C.

Wash and dry the fresh lemongrass stalks (using the bottom part). Bruise them by crushing or gently pounding to release the flavor. Cut into three-inch pieces.

Cut the zest of one lime into thin strips. Avoid the white pith. Slice the other half of the lime into thin slices.

Put the lemongrass stalks, lime zest, and lime slices in a two-cup ovenproof glass measuring cup. Pour in the oil and make sure it covers everything. Place on a pie plate and put in the center of the oven.

Heat for at least an hour or until the temperature of the oil reaches 250°F/120°C. Use a digital thermometer to test.

Remove from the oven, let cool, and pour into a clean container. Store in the refrigerator as is, or strain through cheesecloth and refrigerate.

Grilled Lamb Chops in Rosemary, Lemon, and Thyme Oil

$1/3$ cup white wine

$1/4$ cup **rosemary, lemon, and thyme oil**

1 clove garlic, crushed

$1/2$ tsp rosemary leaves, chopped

$1/2$ tsp thyme leaves, chopped

4 portions lamb chops

salt and freshly ground black pepper

1 sprig rosemary, to garnish

zest of 1 lemon, to garnish

Serves 4

This dish looks as impressive as it tastes, and the addition of the oil makes it something very special.

Make the marinade by whisking together the wine, the rosemary, lemon, and thyme oil, the garlic, and the rosemary and thyme leaves. Pour over the lamb chops, making sure each piece gets covered, cover the dish with plastic wrap, and refrigerate for up to three hours.

Remove the chops from the refrigerator and let them sit at room temperature for about 30 minutes before grilling.

Heat the grill to medium-high. Sprinkle salt and pepper over the lamb chops and grill on both sides until cooked to preference.

Serve with a garnish of rosemary leaves and lemon zest.

Roasted Chicken with Rosemary, Lemon, and Thyme Oil Rub

4 tbsp **rosemary, lemon, and thyme oil**

1 roasting chicken
(5-6 lbs/2.25-2.7 kg)

1 large lemon, chopped

1 sprig rosemary

3 cloves garlic

1 cup coarse salt

salt and freshly ground black pepper

Serves 4-6

You might want to make this just for the wonderful aroma that will fill your kitchen when it is roasting.

Brush the rosemary, lemon, and thyme oil all over the chicken, cover, and marinate for several hours or overnight.

Preheat the oven to 400°F/200°C. Fill the cavity of the chicken with the chopped lemon pieces, rosemary sprig, and garlic cloves. Place on a baking pan, surround the chicken with the coarse salt, and roast for 45 minutes to 1 hour. Check that the chicken is cooked by piercing the thickest part of the leg with a sharp knife or skewer and making sure that the juices run clear.

Remove from the oven, cut the chicken into quarters, and serve.

Sizzling Lemongrass Beef with Asparagus and Red Pepper

15 spears asparagus

2 tbsp **lemongrass and lime oil**

7 oz/200 g bean sprouts

1 red bell pepper, thinly sliced

1 tbsp chopped garlic

1 lb/450 g beef tenderloin, thinly sliced

$1/3$ cup dry white wine

salt and freshly ground black pepper

zest of 1 lime

Serves 4

The combination of flavors and colors in this dish makes for a taste sensation—try adding different vegetables if these are not to your liking.

Clean the asparagus by clean the asparagus by scraping away any woody parts and trimming off the cut end, if necessary. and scraping away any woody parts and trimming off the cut end, if necessary. Slice each spear on the diagonal into thirds. Bring a pot of water to boil and quickly blanch the asparagus. Plunge into ice water.

Heat a wok over high heat and add the lemongrass and lime oil. Add the sprouts, red bell pepper, garlic, and beef and stir-fry for one minute. Add the wine, asparagus, salt, and pepper and stir-fry until the beef is done. Add the lime zest, stir for another minute, and remove from the heat. Place on a serving platter or plates and serve immediately.

Steamed Sea Bass with Lemongrass and Lime Oil

1¹/₂–2 lbs/675–900 g whole sea bass, cut open along one side

1 stalk lemongrass (bottom 6 inches only), thinly sliced

2 cloves garlic, thinly sliced

3 lime slices

salt and freshly ground black pepper

3 tbsp **lemongrass and lime oil**

tomatoes, chopped, to garnish

red, yellow, and green bell peppers, diced, to garnish

Serves 1

Serve this fish with cooked new potatoes on the side, and drizzle the potatoes with a little lemongrass and lime oil for something extra special.

Preheat the oven to 425°F/220°C.

Open up the fish and place the lemongrass, garlic, and lime slices inside. Sprinkle with salt, pepper, and a little of the lemongrass and lime oil. Close and make 3 diagonal slits through the skin along the length of the fish on both sides. Brush the outside with the oil and sprinkle on more salt and pepper.

Lay out a piece of aluminum foil large enough to wrap around the fish. On top of the foil, place a large sheet of parchment paper and place the fish on it. Make an envelope out of the parchment paper, then fold up the aluminum foil around it (doing so will ensure that the fish is steamed). Place on a baking sheet and put in the oven. Bake for 20 minutes.

Remove from the oven, take the fish out of its baking envelope, and place on a serving platter. For a colorful garnish, chop up fresh tomatoes and red, yellow, and green bell peppers and serve on top of the fish.

Walnut Oil

This may be one of the most sumptuous oils. In stores, it is very expensive, so make your own and enjoy it on salads, breads, fruit, etc.

1/2 cup walnuts

1 cup canola oil

Break up the walnuts and dry roast in a skillet over medium heat for about two minutes or until they start to give off a fragrant aroma.

Using a microwave-proof container, heat the oil in a microwave on medium-high for one to two minutes.

Combine 1/4 cup of the heated oil with the walnuts, and process in a blender or food processor until the walnuts are finely chopped. Combine with the remaining oil, pour into a clean jar, and cover. You may leave the oil out for a day so that it absorbs the nutty flavor, then place it in the refrigerator.

Before using, remove the oil from the refrigerator, bring it to room temperature, and strain through cheesecloth. Return to the bottle and use. You can use any remaining chopped nuts as part of the dish you are making.

Orange Spice Oil

Prepare yourself for the most wonderful aroma when this oil is cooking. Try it on sweet breads, rice, fish—the possibilities are endless!

grated peel of 1 orange

20 whole cloves

2 whole bay leaves

1 tsp cinnamon

$\frac{1}{2}$ tsp whole black peppercorns

$\frac{1}{4}$ tsp ground nutmeg

2 tbsp orange juice

1 cup canola oil

Preheat the oven to 300°F/150°C.

Using a mortar and pestle, pound together the grated orange peel, cloves, bay leaves, cinnamon, peppercorns, and nutmeg. Add the orange juice. Mix until it forms a paste.

Transfer the paste to an ovenproof measuring cup and add the canola oil. Set the cup on a pie plate in the center of the oven and cook for about one hour or until the oil reaches 250°F/120°C. Use a digital thermometer to measure.

Remove from the oven, allow to cool, and strain out the flavorings using cheesecloth or a fine-mesh sieve. Pour the strained oil into a clean jar, cover, and refrigerate.

Endive Salad with Roquefort, Figs, and Walnut Oil

6 medium endives

7 oz/200 g Roquefort cheese, crumbled

2 fresh (or dried) figs, chopped

$1/_2$ cup chopped walnuts

$1/_2$ cup **walnut oil**

$1/_4$ cup **fig and spice vinegar**

Serves 6

To make this dish truly special, you will need the addition of the fig and spice vinegar, the recipe for which can be found on page 88. Alternatively you can use ordinary balsamic vinegar.

Cut off the core of the endive and separate the leaves. Put them in a bowl and add the cheese, figs, and walnut pieces.

Combine the walnut oil with the fig and spice vinegar and whisk well. Pour over the salad ingredients and toss gently.

Divide the salad among six plates, and serve.

Banana Walnut Bread

1 cup **walnut oil**

1 cup sugar

2 eggs

3 bananas, very ripe

2 cups all-purpose flour

1 tsp baking soda

1 tsp baking powder

$\frac{1}{2}$ tsp salt

3 tbsp milk

2 tbsp yogurt

$\frac{1}{2}$ tsp vanilla extract

$\frac{1}{2}$ cup walnuts, crushed

Makes 1 loaf

Banana bread is loved by children and adults alike, and the addition of the walnut oil makes for a superior taste.

Preheat the oven to 350°F/175°C.

Grease a 12-inch loaf pan. In an electric mixer, beat the oil and sugar. Beat in the eggs and mashed bananas.

Combine the dry ingredients, then add to the banana mixture. Add the milk, yogurt, and vanilla; beat until blended. Stir in the crushed walnuts and pour the batter into the greased pan.

Bake for 50 minutes to one hour or until a skewer inserted into the middle comes out clean.

Orange Spice Muffins

2 large eggs

²/₃ cup sugar

1 tsp vanilla extract

¹/₂ tsp grated lemon zest

pinch of salt

1 cup all-purpose flour

¹/₂ cup **orange spice oil**

confectioners' sugar, for dusting

orange zest, to decorate

Makes 12

These lovely muffins make a perfect treat for breakfast, brunch, or an afternoon snack.

Preheat the oven to 375°F/190°C.

Using an electric mixer, beat the eggs and sugar together. Beat in the vanilla, grated lemon zest, and salt. Add the flour and mix well. Slowly add in a stream of the orange spice oil and beat until just blended.

Grease and flour a muffin tin. Spoon the mixture into each indentation, filling it about three-quarters full. Place on the middle rack of the oven and bake for about 30 minutes or until a skewer inserted into the middle comes out clean. Remove from the oven, cool, and drop out of the pan. Dust the muffins with confectioners' sugar and add the orange zest to decorate.

Introduction to Vinegars

Our ancestors were very wise about vinegar. The ancient Babylonians used it as a preservative, the Romans drank it, Helen of Troy bathed in it, and Cleopatra dissolved a pearl in it to prove that she could consume a fortune in a single meal. It is mentioned numerous times in the Bible, and Hippocrates recommended its therapeutic properties.

As with oils, there are many types of vinegars: balsamic (produced from selected grapes and fermented for a long time), red and white wine vinegar, rice vinegar, and cider vinegar. There are also specialty vinegars like raspberry, sherry, and champagne vinegar.

Good vinegar can be utilized in so many ways. It is used to prepare and preserve food. It also has health-giving properties: a tablespoon a day of cider vinegar is said to be most beneficial. Even its hygienic properties are well-known; your grandmother was right when she told you that there is nothing like a spray of vinegar for cleaning.

Best of all, vinegar plays an important role in dressings, sauces, and mustards. It adds just the right kick to a barbecue sauce; it brings tanginess to a vinaigrette; it sweetens a wine reduction sauce. Infusing vinegar with other flavors is a great way to add yet another layer of aroma and flavor.

Because vinegar is high in acid, it does not allow botulism bacteria to flourish. Flavored vinegars can be stored in a cool, dark place at first. To prevent mold, make sure that the additions are totally submerged. It is a good idea to store vinegar in the refrigerator after a week or two.

The following points are important to remember:

- Use non-reactive containers (glass, enamel, or stainless steel) that have been thoroughly cleaned, but not by metal scouring pads.
- Use good quality vinegar, fresh herbs that have been washed and dried, and spices that are not stale.
- Heating the vinegar and bruising the herbs help speed the infusion process.
- Strain out solids when the vinegar has reached the flavor that you desire.
- Do not store for extended periods of time.

Dill and Peppercorn Vinegar

The classic combination of dill and peppercorns makes for a very tasty vinegar dressing.

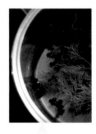

6 sprigs fresh dill

1 cup cider vinegar

1 tsp whole black peppercorns

Wash and dry the dill.

In a saucepan over medium heat, bring the vinegar to a boil. Lower the heat and simmer for 2 minutes. Add the dill and peppercorns, turn off the heat, and let sit for several minutes until cooled.

Pour into a clean jar, seal and keep in a dark place until ready to use or refrigerate.

Note: A few pieces of chopped fresh mint can also be added to the vinegar as an interesting additional flavor.

Garlic, Chili, and Red Wine Vinegar

If ever a vinegar came with a kick, this is it!

4 cloves garlic

$\frac{1}{2}$-inch piece red hot chile

1 cup red wine vinegar

Peel the garlic and slice lengthwise into quarters. Wearing gloves to prevent direct contact with the chile, chop and deseed. Put the garlic slices and chile pieces into a clean jar.

    Heat the vinegar in a small saucepan over medium heat until it starts to bubble around the edges. Remove from the heat and pour the vinegar into the jar with the garlic and pepper. Let it cool, then cover and store. The vinegar can be left in a dark place or refrigerated.

Dill and Peppercorn Vinegar

Cucumber Salad

4-6 cucumbers

$1/2$ red onion, thinly sliced into
half rings

$1/8$ cup finely chopped dill

$1/2$ cup **dill and peppercorn vinegar**

2 tbsp sugar

salt and white pepper

Serves 4-6

This refreshing salad is great for a picnic on a summer's day. Try serving it with cold poached salmon.

Wash and dry the cucumbers and score them lengthwise with a fork. (You can also use a vegetable peeler and leave thin strips of peel on.) Slice into very thin slices. Combine with the onions and dill.

Whisk together the vinegar, sugar, and salt and pepper to taste. Pour over the cucumber mixture, toss well, and cover. Refrigerate and allow to marinate for several hours.

Garlic, Chili, and Red Wine Vinegar

Garlic and Chili Barbecue Marinade

One 12 oz/340 g jar preserved apricots

$^1/_3$ cup **garlic, chili, and red wine vinegar**

$^1/_4$ cup brown sugar

2 cloves garlic, crushed

2 tbsp Dijon mustard

1 tsp fresh ginger, grated

Makes about two cups

This barbecue marinade is perfect on almost anything. Try it on ribs, chops, or chicken and transform your barbecue food!

Combine all the marinade ingredients in a blender and process until smooth. Cover and refrigerate.

Note: If using the marinade for chicken pieces or spareribs, it is best to marinate them overnight and then head straight to the barbecue!

Garlic, Chili, and Red Wine Vinegar

Spicy Icy Gazpacho

$^1/_2$ onion, coarsely chopped

2 cloves garlic, crushed

3 tbsp olive oil

$^1/_4$ cup **garlic, chili, and red wine vinegar**

2 lbs/900 g tomatoes

1 large cucumber, peeled and chopped

1 green bell pepper, coarsely chopped

$^1/_2$ stalk celery, sliced

$^1/_4$ cup fresh cilantro leaves

$^1/_8$ cup tomato paste

$^1/_2$ cup tomato juice, for thinning (optional)

salt and freshly ground black pepper

parsley and cilantro oil ice cubes (optional)

Serves 6-8

A twist on the Spanish classic, the addition of the garlic, chili, and red wine vinegar makes this even better than the original!

Combine all the ingredients except the tomato juice and salt and pepper in a food processor and process until a chunky puree forms. Add tomato juice if a thinner consistency is desired. Season with salt and pepper to taste.

Pour into a container, cover, and refrigerate for several hours. Remove from the refrigerator immediately before serving.

Note: If you have parsley and cilantro oil on hand, freeze some in ice cube trays and float three in each bowl of gazpacho.

Caraway and Cider Vinegar

A drizzle of this flavorful vinegar goes a long way. It is especially good for all kinds of fresh and cooked cabbage dishes.

1 tbsp caraway seeds

1 cup cider vinegar

In a dry skillet, heat the caraway seeds for one to two minutes.

In a saucepan over medium heat, heat the vinegar until it starts to bubble around the edges of the pan.

Add the toasted caraway seeds to the vinegar, pour into a clean jar, cover, and store in a dark place.

Mango, Tarragon, and Champagne Vinegar

Champagne is definitely not just for drinking on special occasions. Try this recipe using champagne vinegar and make your own special occasion!

4–5 sprigs tarragon

1 cup champagne vinegar

$\frac{1}{2}$ cup diced mango

$\frac{1}{4}$ cup orange juice

$\frac{1}{4}$ cup sugar

$\frac{1}{4}$ tsp vanilla

dash of salt

Chop the tarragon leaves and combine with all the other ingredients in a saucepan. Bring to a boil over medium-high heat and let simmer for 1-2 minutes.

Remove from the heat, let cool, and transfer to a blender. Purée until smooth. The vinegar can be strained and stored, or stored as is and strained before using. Keep covered in a clean jar in a dark place or refrigerate

Beet and Caraway Borscht

3 tbsp olive oil

1 tbsp caraway seeds

2 medium onions, chopped

3 cloves garlic, crushed

1 potato, peeled and diced

6 cups vegetable stock

$1/4$ cup **caraway and cider vinegar**

2 lbs/900 g cooked beets, diced

salt and freshly ground black pepper

1 tbsp lemon juice

$1/2$ cup chopped parsley and cilantro, plus extra to garnish

sour cream, to serve

Serves 4

The color of this soup is lovely and the taste is exquisite!

Heat a large pot over medium-high heat, add the olive oil, and sauté the caraway seeds for 1 minute. Add the onion and stir until it becomes translucent. Add the garlic and stir in.

Add the potato to the pot, sauté for a few minutes, and add the stock and vinegar. Bring to a boil over high heat. Lower the heat and simmer for about 8 minutes. Add the cooked beets, salt, and pepper and bring to a boil again. Lower the heat, partly cover the pot, and simmer for 30 to 45 minutes or until the potatoes are cooked through.

Add the lemon juice and herbs and cook for 5 more minutes. Taste and adjust the seasonings.

When the soup cools a little, remove one-third of it to a blender or food processor and process until puréed. Add back to the pot, mix, reheat, and serve.

Serve with a dollop of sour cream and additional chopped herbs.

Caraway and Cider Vinegar

Braised Red Cabbage and Apples

1 tbsp whole caraway seeds

1 tbsp vegetable oil

1 red onion, halved and thinly sliced

2 tbsp brown sugar

1 small red cabbage, shredded

2 apples, peeled and thinly sliced

2 tbsp red wine

$^1/_2$ cup apple juice

$^1/_8$ cup **caraway and cider vinegar**

salt and freshly ground black pepper

1 tsp lemon juice

Serves 6

Red cabbage absorbs new flavors extremely well, and the addition of the vinegar makes this a delicious side dish.

In a saucepan over medium heat, dry roast the caraway seeds for about 1 minute until they start to give off an aroma.

Heat the oil in a large pot over medium heat, add the onion, and sauté for 5 minutes until it becomes translucent. Add the brown sugar, stir, and add the cabbage and apples. Stir for a few minutes until the cabbage wilts. Add in the red wine, apple juice, and vinegar. Add the toasted caraway seeds and salt and pepper to taste. Bring the mixture to a boil, lower to a simmer, add the lemon juice, cover, and cook for 30 minutes.

Note: Braised cabbage is wonderful served with chicken, meat, or pork dishes.

Mango, Bell Pepper, and Pineapple Salsa

1 tbsp olive oil

1 tsp brown sugar

1 mango, peeled and diced

$\frac{1}{2}$ cup pineapple chunks

$\frac{1}{2}$ red bell pepper, diced

$\frac{1}{2}$ yellow bell pepper, diced

$\frac{1}{2}$ green bell pepper, diced

$\frac{1}{8}$ cup **mango, tarragon, and champagne vinegar**

salt and freshly ground black pepper

$\frac{1}{8}$ cup chopped tarragon leaves

Makes about two cups

This salsa is so colorful and tasty that you will look for dishes to serve it with: grilled swordfish, steamed bass, and even barbecued chicken all complement it perfectly.

Heat the olive oil in a skillet over high heat. Add the brown sugar and stir.

As soon as the sugar dissolves, add the mango, pineapple, and bell peppers. Keep stirring over high heat and slowly pour in the mango, tarragon, and champagne vinegar. Add the salt, pepper, and tarragon leaves, and stir for another minute or two. Keep an eye on this—you want the ingredients to just slightly char but not really cook. Remove from the heat and serve warm.

Lemongrass, Ginger, and Garlic Vinegar

This is a wonderfully exotic vinegar, inspired by Southeast Asian cuisine.

2 stalks lemongrass

3 cloves garlic, peeled

1 tbsp grated ginger

1 cup rice wine vinegar

Wash and dry the lower portion of the lemongrass stalks, then crush or bruise them slightly. Cut them if you would like smaller pieces in the jar. Cut the garlic cloves in half lengthwise.

Place the lemongrass, garlic, and ginger pieces into a clean jar.

In a saucepan over medium heat, heat the rice wine vinegar until it starts to bubble around the edges of the pan. Remove from the heat, let cool a bit, then add to the jar with the other ingredients. When completely cool, cover the jar and store in a dark, dry place.

Note: If rice wine (a staple of Asian cuisine) is not available, use cider vinegar or plain white vinegar.

Ginger and Star Anise Vinegar

Try this vinegar on a salad, or even better in chutney, as these three ingredients are simply delicious when combined.

1 cup red wine vinegar

3 whole star anise

1 ½-inch/3 ½-cm piece fresh ginger, sliced

Bring the red wine vinegar to a boil in a saucepan over medium heat. Let simmer for two minutes. Remove from the heat, add the ginger and star anise and let cool.

Pour the mixture into a clean jar. Cover and keep in a dark place or refrigerate. Test for taste and filter out the star anise and ginger when ready to use.

Lemongrass, Ginger, and Garlic Vinegar

Asian Coleslaw

¹/₄ cup **lemongrass, ginger, and garlic vinegar**

2 tbsp sugar

¹/₄ tsp salt

¹/₂ cup mayonnaise

¹/₂ head large green cabbage, shredded

2 carrots, grated

¹/₄ cup crushed cashews

Serves 4-6

Once you have made this mouthwatering variation of traditional coleslaw, you will find it difficult to go back to the original!

To make the marinade, combine the lemongrass, ginger, and garlic vinegar and sugar in a small saucepan and stir over medium heat until the sugar is dissolved. Remove from the heat and add the salt and mayonnaise.

Combine the shredded cabbage and carrots in large bowl. Add the marinade and toss well. Transfer to a serving bowl, scatter the cashews on top, and serve.

Lemongrass, Ginger, and Garlic Vinegar

Skewered Shrimp with Lemongrass Yogurt Dip

Shrimp

4 stalks lemongrass

24 jumbo shrimp

coarse salt and freshly ground pepper, to garnish

Dip

³/₄ cup plain yogurt

2 tbsp **lemongrass, ginger, and garlic vinegar**

1 cucumber, peeled and diced

1 clove garlic, crushed

¹/₂ tsp salt

dash of white pepper

¹/₂ lemon, juiced

¹/₂ tsp grated fresh ginger

Serves 4

This dish is lots of fun to serve at a party—your guests will love having the shrimp served on lemongrass skewers, and the dip is absolutely delicious.

To make the shrimp, cut each lemongrass stalk in half lengthwise and trim to 12-inch-long pieces. Using a sharp knife, whittle away the end of the lemongrass stalk to form a pointy end.

Peel and devein the shrimp. Thread 3 shrimp onto each lemongrass stalk by carefully inserting the pointy end of the stalk into one side of the thick part of shrimp and then out the other side. Bend the shrimp a little and thread the stalk again into the less meaty part. Repeat the process so that each lemongrass stalk has 3 shrimp that have been pierced twice.

Cook the shrimp on a hot grill for about two minutes on each side—just until they turn pink.

To make the dip combine the dip ingredients, mix well, and transfer to a

serving bowl. If not using immediately, cover with plastic wrap and store in the refrigerator.

Arrange the shrimp skewers on a platter and sprinkle with coarse salt and freshly ground black pepper. Serve with a bowl of the lemongrass yogurt dip.

Ginger and Star Anise Vinegar

Apricot and Fig Chutney

$^1/_2$ cup **ginger and star anise vinegar**

$^1/_2$ cup apple cider

$^1/_2$ cup water

1 red onion, coarsely chopped

$^1/_4$ cup brown sugar

$1^1/_2$-inch/$3^1/_2$-cm piece fresh ginger, sliced

$^1/_8$ tsp black peppercorns, slightly crushed

1 cup dried apricots (about 20), chopped

6 dried figs, stems removed and chopped

Makes about 1$^1/_2$ cups

Try this recipe with dried fruit or fresh fruit—whichever you prefer. Chutneys are wonderful and they make a great accompaniment to turkey or meat dishes.

Combine the ginger and star anise vinegar, cider and water in a saucepan over medium-high heat. Add the onion, cook for two minutes, and add the brown sugar. Continue to stir over medium heat and add the remaining ingredients. Bring to a boil, lower the heat, and simmer until all the liquid is absorbed. Add more boiling water if it is used up and the fruit is still not tender. You can add one star anise from the vinegar, but remove it after cooking.

Adjust the seasonings to taste, let cool, transfer to a bowl with a lid, and refrigerate.

Rosemary and Garlic Balsamic Vinegar

The combination of rosemary and garlic makes this one of the most fragrant vinegars, and it tastes every bit as good as it smells!

Ten 2-inch sprigs rosemary

4 cloves garlic

1 cup balsamic vinegar

Wash the rosemary sprigs, dry, and tear off the leaves from the stems. Split the garlic cloves in half lengthwise. Combine the leaves and garlic halves in a clean jar.

In a saucepan over medium heat, heat the balsamic vinegar until it just starts to bubble around the edges of the pan. Wait until it cools a little, then pour into the jar with the rosemary and garlic. When it is completely cool, cover the jar and store in a cool, dark place. Check occasionally to see whether the vinegar has reached the desired strength.

Before using, strain the vinegar through a fine sieve or cheesecloth into clean jars. Add a fresh sprig of rosemary for decoration and again cover and store in a cool, dark place.

Basil, Chive, and Lemon Vinegar

This fresh-tasting vinegar is very useful for transforming salads when you have unexpected guests.

zest of ¹⁄₂ lemon

5 basil leaves

10 stalks chives

1 cup white wine vinegar

When peeling the lemon for the zest
(using only half the lemon), be sure to avoid the white pith.
Wash and dry the basil leaves and the chives, then crush
them or chop coarsely. Place the zest, basil, and chives in a
clean jar.

   In a saucepan over medium heat, heat the white wine
vinegar until it starts to bubble around the edges of the
pan. Wait until it cools just a little, then add it to the jar with
the other ingredients. When it is completely cool, cover the
jar and store in a cool, dark, and dry place.

Grilled Portobello Mushrooms with Rosemary Garlic Drizzle

8 fresh Portobello mushrooms

2 tbsp olive oil

salt and freshly ground black pepper

⅓ cup **rosemary and garlic balsamic vinegar**

fresh Parmesan cheese

rosemary sprigs, to garnish

Serves 2-4

There is something very special about the flavor of Portobello mushrooms combined with balsamic vinegar. Add rosemary and garlic, and it is pure heaven!

Preheat the broiler. Clean the mushrooms and remove the stems. Place the mushrooms on a baking sheet and brush both sides with olive oil. Sprinkle with salt and pepper. Broil for about three minutes. Turn over and broil the other side.

While the mushrooms are broiling, pour the rosemary and garlic balsamic vinegar into a saucepan and bring to a boil. Lower the heat and simmer for several minutes until the mixture starts to reduce.

When the mushrooms are soft, remove from the broiler, place on a serving platter, and drizzle over the reduced vinegar. Grate the Parmesan on top, decorate with a rosemary sprig, and serve immediately.

Marinated Rosemary and Garlic Chicken Breasts

¹/₄ cup **rosemary and garlic balsamic vinegar**

¹/₂ cup olive oil

4 cloves garlic, crushed

salt and freshly ground black pepper

2 lbs/900 g chicken breasts, boned and skinned

rosemary sprigs, to garnish

Serves 4-6

This recipe is really simple and truly delicious. It keeps several days in the refrigerator, so make a batch and grill the chicken whenever you feel like it.

To make the marinade, combine the rosemary and garlic balsamic vinegar, oil, and garlic and mix well. Season with salt and pepper to taste.

Pound the chicken breasts until thin and place in the marinade. Cover and refrigerate at least overnight.

When ready to cook, remove the breasts from the marinade and season with salt and pepper. Heat a skillet or grill pan over high heat, add the chicken breasts, and sauté on both sides until cooked through. Discard the remaining marinade—do not use it again since the raw chicken was marinating in it.

Serve the chicken breasts immediately, garnished with fresh rosemary sprigs.

Warm Potato and Artichoke Salad with Basil, Chive, and Lemon Dressing

1-1$^1/_2$ lbs/450-675 g small potatoes

5 artichoke bottoms, cooked

$^1/_8$ cup chopped pickles

1 tbsp fresh dill, chopped

20 stalks chives, chopped

4 tbsp **basil, chive, and lemon vinegar**

1 tsp Dijon mustard

2 tbsp olive oil

1 tbsp fresh lemon juice

salt and freshly ground black pepper

Serves 4-6

This is a great summer salad—the potatoes and artichokes go beautifully with the basil, chive, and lemon dressing.

Clean the potatoes, leaving the skins on, and cook in boiling water until soft. Cut into bite-size pieces. Cut the cooked artichoke bottoms into bite-size pieces and combine in a mixing bowl with the cooked potatoes. Add the pickles, dill, and chives.

   Whisk together the basil, chive, and lemon vinegar, mustard, olive oil, and lemon juice. Season with salt and pepper to taste. Pour over the potato and artichoke mixture and mix.

   Serve immediately or store covered in the refrigerator and bring to room temperature before serving.

Fig and Spice Vinegar

This vinegar is perfect for spicing up winter dishes—the distinctive aroma of the cinnamon and cloves will remind you of all things festive.

6 whole dried figs, stems removed, quartered

³/₄ cup cider vinegar

¹/₄ cup balsamic vinegar

2 tbsp sugar

¹/₂ tsp cinnamon

1 tsp ground cloves

¹/₄ tsp vanilla

Cut the figs into quarters.

Add all the ingredients to a saucepan and bring to a slow boil over medium heat. Let simmer for two minutes. Remove from the heat, let cool, pour into a clean jar, and cover. Store in a dark, cool place.

Pomegranate and Vanilla Vinegar

The combination of the pomegranate seeds and the vanilla extract make a glorious color. When stored in a pretty glass jar, this will look lovely on display in your kitchen.

1 cup white wine vinegar

$1/3$ cup pomegranate seeds

1 tsp vanilla extract

Heat the white wine vinegar in a saucepan over medium heat. Bring to a boil and let simmer for two minutes.

Pour the pomegranate seeds and vanilla into a clean jar. When the vinegar has cooled a little, add it to the jar. Cover and filter out the seeds within two weeks. Store in a dark place or refrigerate.

Baked Cheese and Spiced Fig Puffs

1 lb/450 g frozen puff pastry sheet (1 sheet)

all-purpose flour, for dusting

4 oz/115 g cream cheese

2 oz/55 g Roquefort

2 oz/55 g goat cheese

¼ cup chopped pistachios

1 egg, beaten

2 tbsp **fig and spice vinegar**

Serves 6

Your guests will love this appetizer. The puffs look good, taste delicious, and are perfect for the start of a wonderful party.

For the pastry puffs, defrost the pastry sheet in the refrigerator (preferably overnight).

Preheat the oven to 400°F. Grease 2 x 12-cup muffin tins.

Prepare a floured work surface and lay out the pastry sheet on top of it. Roll out to a thickness of ⅛ inch/3 mm. Cut out circles of dough slightly larger than the circumference of the holes in a muffin tin. Place each circle of dough into each muffin hole. Spread out to cover the bottom and a little way up the sides. Brush the surface with the egg wash.

Place in the oven and bake for about 12 minutes—until the pastry puffs begin to turn golden. Remove from the oven, let cool, and remove the pastry "muffins" from pans.

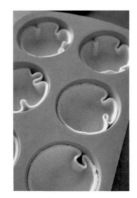

When ready to serve, fill each pastry muffin with a teaspoonful (or more) of the cheese mixture. Top with a small piece of fig and drizzle some of the fig vinegar over it. Put on a baking sheet, place in the oven heated to 400°F, and bake for 1 to 2 minutes—until the cheese melts. Remove from the oven and serve.

For the filling, combine the cheeses and chopped pistachios in food processor. Process until well blended. Store covered in the refrigerator until ready to serve.

When ready to prepare the pastry puffs for serving, take the fig quarters out of the vinegar and cut into smaller pieces. Place one small piece on top of each mound of cheese.

Salmon Carpaccio with Pomegranate Seeds and Basil

1 lb/450 g salmon fillet, skinned and sliced paper thin

$^{1}/_{2}$ cup olive oil

3 tbsp **pomegranate and vanilla vinegar**

2 tbsp freshly squeezed lemon juice

salt and freshly ground black pepper

$^{1}/_{4}$ cup fresh basil leaves, chopped

zest of $^{1}/_{2}$ lemon, cut into thin strips

$^{1}/_{3}$ cup fresh pomegranate seeds

Serves 6

This is a perfect dish for a weekend brunch. It looks beautiful and takes no time at all to prepare!

Arrange the salmon slices on a serving platter large enough to accommodate all of them without overlapping.

Whisk together the olive oil, pomegranate and vanilla vinegar, and lemon juice. Season to taste with salt and pepper and pour over the salmon slices. Cover with plastic wrap and refrigerate for at least 30 minutes.

Remove the plastic wrap from the serving platter, and sprinkle the salmon with the basil leaves, strips of lemon zest, and pomegranate seeds.

Note: This recipe calls for raw salmon, but you can easily substitute smoked salmon.

Pomegranate and Vanilla Vinegar

Sauteed Pears with Whipped Yogurt and Pomegranate Cassis Reduction

3 tbsp **pomegranate and vanilla vinegar**

2 tbsp crème de cassis

2 tsp maple syrup

1/2 cup plus 1 tablespoon heavy whipping cream

freshly ground black pepper

2 pears, peeled and thinly sliced

1 tsp sugar

1/3 cup whole plain yogurt

fresh mint leaves, to garnish

Serves 4

This dish not only sounds sophisticated, but it tastes very extravagant—perfect for when you want to impress!

In a small skillet over medium-high heat, combine the pomegranate and vanilla vinegar, crème de cassis, and maple syrup. Keep stirring over the heat as the mixture reduces and becomes syrupy. After about four minutes, add the tablespoon of cream and a dash of black pepper. Lower the heat, keep stirring, and right before you serve it, add the sliced pears and sauté for another minute. The pears should be nicely coated, but keep an eye on them. Make sure the syrup does not burn.

In a mixing bowl, beat the 1/2 cup whipping cream until it starts to thicken, then add the sugar and keep whipping. When peaks form, stop beating and fold in the yogurt.

Remove the pears from the pan and place on serving plates. Top with the whipped cream and yogurt mixture, drizzle a bit of the pomegranate syrup over them, and serve.

Garnish with fresh mint leaves.

Index